SCIENCE WITHOUT SENSE

The Risky Business
Of Public Health Research

by
Steven Milloy

Copyright © 1995 by Steven J. Milloy.
Second printing: September 1997.

All rights reserved.

First edition. Published by Cato Institiute,
1000 Massachusetts Ave., N.W.
Washington, D.C. 20001

Library of Congress Catalog Number: 95-72177

International Standard Book Number: 1-882577-34-5

PREFACE

Years ago, as I was about to graduate from The Johns Hopkins University School of Hygiene and Public Health with a hard-earned degree in biostatistics, it occurred to me that a career in public health actually might be somewhat bleak, professionally if not financially.

I realized the average person in this country lives well past 70 years of age. The average lifespan will approach 80 sometime around the year 2010. Exactly how much better can the American public's health get? How much need can there be for public health research professionals in such a healthy society? It's simple supply and demand.

Research professionals, I thought to myself, will become the public health equivalent of the Maytag repairman. No one will need us anymore. So I graduated, went on to law school and never looked back. But as it turned out, I grossly underestimated the initiative of the entrepreneurs in our public health community.

In fact, there's something of a gold rush going on in public health today. Thanks to the general public's neuroses about health, some strategic fearmongering, and, of course, political considerations, public health has struck it rich — to the tune of billions of dollars in annual revenues.

Who would have thought it possible? It's the irony of ironies. More than half of us can expect to live past

75 years of age. Yet there are more public health professionals finding more public health problems than ever before!

I'm sure there are many people out there who want to know exactly how to take advantage of this curious situation. That's what this little guide is all about.

INTRODUCTION

Want to get ahead in public health? Want to "discover" a health risk that will make you famous? Tired of "science" and its cumbersome "scientific method"? Frustrated with customary practices and limitations of traditional public health disciplines like epidemiology and biostatistics?

If you answered "yes" to these questions, then risk assessment is for you. But not just any type of risk assessment. We're not talking about science-driven, common sense-type risk assessment. That type of risk assessment too often gets soft-peddled or even ignored. Your good work — and career — end up in public health oblivion.

What we're talking about is the kind of risk assessment that's not confined by the restricting chains of real science. Risk assessment that knows no shame, that will stoop to any level to achieve its self-fulfilling prophecies. In other words — assess for success!

This is an unabashed guide to using risk assessment to climb the public health career ladder. It outlines the E-Z approach to "discovering," "proving" and "marketing" health risks. Here you'll learn how to pick the "right" risk, how to do the "science," how to defend your work, how to "set up" peer review. We'll even tell you which professional journals are best for your work.

This guide has everything you need to know about how to create a risk that will electrify the public, launch you into the pantheon of public health and land those big fat research grants from the federal government.

This is THE guide for the public health superstar wanna-be. Forget about science, the scientific method and all that other junk you learned in college and graduate school. Frankly, that stuff just doesn't cut it any more. This is what you need today.

CHAPTER 1
Picking the Right Risk

Finding the right risk to "discover" is the critical first step. If you pick the right risk, its intrinsic characteristics will make most of the risk assessment process a mere formality. Pick the wrong risk and the only thing at risk is your career.

The risk should be unprovable

The very existence of your risk must be unprovable by conventional scientific methods. After all, if it was provable, somebody else (like a real scientist), would already have done the work and your risk assessment wouldn't be necessary. A risk may be unprovable either because it doesn't actually exist or because the risk is too small to evaluate with science. In either case, fortunately for you, it's technically impossible to disprove such a risk.

For example, consider Superfund, the federal program to clean up hazardous waste sites. Sites are designated for clean-up where it is calculated that someone's chance of getting cancer from the site is 1 in 10,000 or more. This risk is so small that it could never be scientifically shown to exist. It would take a study with at least 500 million subjects — about two times the current U.S. population — to prove such a small risk exists. Even a 1 in 1,000 risk would require a study with five million subjects! (A typical study contains just a few hundred subjects; on rare occasions, a few thousand.) Using an unprovable risk offers several advantages.

First, you can never be proved wrong. *This is very important.* Of course, you can never be proved right either, but that's a small detail, one that really doesn't matter in the grand scheme of things. You just need to allege, not prove.

Second, an unprovable risk allows you to make outrageous assumptions about your risk, including the threshold assumption that your risk exists at all.

Some of the more famous unprovable cancer risks are dioxin, electromagnetic fields, hazardous waste sites, environmental tobacco smoke, household radon, chlorinated drinking water and pesticide residue in foods.

The risk should be ubiquitous

Your risk should be one that lots of people, if not everyone, come in contact with or are "exposed" to on a regular basis. But it should be difficult-to-impossible to measure how much exposure there actually is. This allows you to make up how much exposure there is and how to measure it. More about this later.

For example, consider dioxin, which many have tried to link with cancer. Dioxin is a by-product of natural and man-made combustion processes involving chlorine. Natural processes include forest fires, volcanoes and compost heaps. Man-made processes include municipal, hospital and hazardous waste incinerators, internal combustion engines, chemical manufacturing processes and residential wood burning.

As a result of these processes, dioxin is *everywhere*. It's in our air, it's in our food, it's in what we drink. In other words, it's unavoidable. It's virtually impossible to measure exactly how much dioxin we're all exposed to. But it's a lot.

In contrast, specific on-the-job risks or specific infectious diseases are not everywhere. They can be identified more easily and can be avoided by people. But they don't grab the public's attention.

Well-known ubiquitous cancer risks include electromagnetic fields, hazardous waste sites, environmental tobacco smoke, radon, chlorinated drinking water, and pesticides. You'll note that this is the same list of famous "unprovable" risks. This is why they're famous.

The risk should be intuitive to the public

The right risk is one that is logical and intuitive to the general public. Industries that use smokestacks, dump wastewater into rivers or engage in other aesthetically unpleasant activities are sources of good risks. After all, it's obvious that the gunk coming out of the smokestack is harmful. Just think about those overhead electric power lines, you just *know* something bad is coming off them. With that kind of risk, the public doesn't even need to read your research. They knew it all along. You just got around to proving it "scientifically."

Naturally occurring radiation, that is, low levels of radiation from the earth and space (but not the sun!), is one risk that is not intuitive. This type of radiation is unavoidable. Yet public health types would have us believe even naturally occurring radiation puts us at risk of cancer. Well now, let's see if I've got this straight. It's natural. It's unavoidable. So tell me one more time — how is it harmful?

A good risk cannot be defended easily

Whoever is responsible for the risk must be at a public relations disadvantage in defending itself against your accusations. This is easy where you're dealing with "obvious" risks, like those that are intuitive. Other good risks involve things that should be "pure and natural" but have been "contaminated" by man. Food and drinking water with pesticides, those blasts of diesel exhaust you get from buses, indoor air with tobacco smoke, carpet fumes and cleaning agents. If nothing else, no one likes these things regardless of whether they actually are health risks.

Human "vices," where it's easy to take the moral high ground (like smoking and drinking, for example), are also good. Government activities, especially anything to do with radiation or nuclear weapons production, are always good targets. The government can't defend itself; if it tries, your issue gets the added — and very juicy — benefit of being the subject of a government coverup.

Risks should be involuntary

Ideally, your risk should be involuntary. The public perceives these risks to be thrust upon them without their choice or consent. A lawyer would say there's no "assumption of risk." Somebody else is doing something to them. Such risks invoke the outrage factor. Examples here include smokestack industries and electric power lines near homes and schools, pesticide "contamination" of food and water and second-hand tobacco smoke.

In stark contrast, the public tends to get less incensed about risks voluntarily encountered...overeating, driving too fast, smoking and drinking too much. Most of us take our health pretty seriously and we just don't like other people putting us in danger. It's okay if we speed on the interstate or carry around an extra 20 or 30 pounds. But heaven help *someone else* we perceive to be threatening us.

Reducing or eliminating the risk should involve no perceptible personal sacrifice

Don't pick a risk that would require people to sacrifice something near and dear to them. That means fast food, sweets or artificial sweeteners, cellular telephones and the like. Even if there are real risks associated with such things, we value them too much to believe it's worth relinquishing them, even for our health. Recently someone actually tried to associate eating hot dogs with leukemia. Good luck with that one!

Pick risks that others are responsible for. It's the electric power companies that will have to bury those dangerous power lines. It's the chemical companies that have to find substitutes for chlorofluorocarbons (CFCs) threatening the ozone layer. It's the city that will have to pay for cleaning up that unsightly landfill. Somebody else will have to give up smoking indoors. These risks involve no tangible personal sacrifice and are easy for people to get indignant about.

Pick on the unsuspecting

Lastly, pick on a risk that's novel. The public will be surprised; those responsible won't be ready to defend themselves. By the time a defense can be mounted and your study gets the sound thrashing it deserves, your newly won fame will have carried you on to bigger and better things. For instance, remember when the cellular telephone scare was launched on *Larry King Live*?

The next day the stock prices of companies in the cellular telephone industry fell through the floor.

So you've picked your risk. Now what?

It's time to put it all together. The next few of chapters will show you how to use epidemiology, statistics and toxicology to make things happen.

CHAPTER 2
2-4-6-8 We Want to Associate!

Epidemiology is your key to success. Remember, it's the study of real people in the real world... or at least that's the way you should play it. Epidemiology is very convincing to the public — even though it's often no more reliable than a shaky alibi. So you need to be careful.

Epidemiology is the study of, that's right, epidemics. Arising from the Scientific Revolution of the 17th century, epidemiologic studies have been responsible for many genuine advancements in public health. It's how scurvy among 19th Century sailors was linked with vitamin C deficiency, how cholera outbreaks in 19th Century London were associated with untreated drinking water and how typhoid fever was found to be contagious. Epidemiology has a celebrated history, and its distinguished coattails can take you a long way.

There are two basic types of epidemiology studies that you can perform — cohort and case-control. Avoid cohort studies. They involve following a specific group of people into the distant future. Although cohort studies are the better type of epidemiologic study, they can take 20 years or more to complete. You would have to put your ambitions on hold. By the time your results are in, the general public may have wised up and called a halt to the public health gold rush.

On the other hand, case-control studies are preferred because they're fast. Instead of following a group of people into the future, you simply scrounge up a group you can look at in retrospect. It's like Monday morning quarterbacking, only better. At the end of this game, you can adjust the score almost anyway you want.

All you need is a group of people with the disease you're interested in (the cases) and another group of persons without the disease (the controls). Survey the cases and controls (we'll talk about how to do this in a later chapter) to determine who has been exposed to the risk you're studying. If the prevalence of exposure among the cases is greater than the prevalence of exposure among the controls, you may have a winner. For our purposes, we'll call the statistical representation of this comparison the relative risk.

$$\text{Relative risk} = \frac{\text{Prevalence of exposure among cases}}{\text{Prevalence of exposure among controls}}$$

What does relative risk mean? Let's say you've studied the association between high fat diet and lung cancer. You've calculated a relative risk of 6. The correct interpretation of this relative risk is that the incidence of high fat diets in the study population was six times greater among those with persons with lung cancer than those without lung cancer. Now is that boring or what? This interpretation will take you nowhere fast.

You need to reword and generalize this interpretation to give it some sex appeal. A risk assessor on the make would say something like "this study shows the risk of lung cancer is six times greater among persons with high fat diets." Notice how we've replaced "incidence" with "risk," two very different concepts and used the word "shows."

"Incidence" means we merely observed the reported result in our study. "Incidence" does not imply, one way or the other, that a high fat diet is associated with lung cancer. By replacing "incidence" with "risk," however, we communicate that a high fat diet *causes* lung cancer.

Our study didn't really say that, but don't worry. That's a small detail that the general public won't notice. Finally, use of the word "shows" implies the study *proves* the risk. In fact, with a single epidemiologic study, it's impossible to prove anything except the limited observations of that study.

The size of your relative risk is very...no, extremely...no, critically important. The basic rule is simple: the higher the relative risk, the more convincing the association you want to prove.

INTERPRETING RELATIVE RISKS	
Relative risk	**Interpretation (career implications)**
Greater than 3	Strong association (jackpot!)
Between 2 and 3	Weak association (may need life support)
Between 1 and 2	Very weak association (call the coroner)
1	No association (sorry)
Less than 1	Negative association (whoops!)

Now remember, technically a relative risk is only statistical association. It's an ~~apparent~~ relationship between the exposure and disease of interest. Notice the word "apparent" has been struck out. This is not a typo. It's just that you should pretend you never read it. As a matter of science, we really don't know whether the statistical associations identified through epidemiology are real or not. After all, we've only identified them through statistics, and statistics are not science. If science is the sun, statistics are Pluto. In fact, all sorts of wacky associations can be identified through statistics, as shown by the following chart. Tap water and miscarriages, for instance, or whole milk and lung cancer.

SAMPLE STATISTICAL ASSOCIATIONS	
Exposure and disease	**Reported relative risk (by size)**
Environmental tobacco smoke and lung cancer	1.19
Consuming olive oil and breast cancer	1.25
Vasectomy and prostate cancer	1.3
Obesity in women and premature death	1.3
Sedentary job and colon cancer	1.3
3 cups of coffee per week and premature death	1.3
Birth weight of 8+ pounds and breast cancer	1.3
Baldness in men under 55 and heart attack	1.4
Eating margarine everyday and heart disease	1.5
Drinking tap water and miscarriage	1.5
Regular use of mouthwash and mouth cancer	1.5
Abortion and breast cancer	1.5
Eating yogurt and ovarian cancer	2.
Drinking whole milk and lung cancer	2.14
Obesity in nonsmoking women and premature death	2.2
Eating red meat and advanced prostate cancer	2.6
Chlorinated drinking water and bladder cancer	2 to 4
Douching and cervical cancer	4.
Workplace stress and colorectal cancer	5.5
Eating 12+ hot dogs per month and leukemia	9.5
Wearing a brassiere all day and breast cancer	12,500.

Now between you and me, if you start worrying whether associations you identify through epidemiology make sense, you'll never cut it in risk assessment. A well-developed conscience is not necessary here. So for your purposes, you shouldn't really care whether an association is fact or fiction, only that you've found it. But there is this thing called biological plausibility that you will need to remember.

In addition to just finding a statistical association between exposure and disease, you're supposed to show the statistical association is biologically plausible. That is, it should make sense from a biological standpoint. For example, it is generally accepted as biologically plausible that too much exposure to the sun's ultraviolet rays is associated with an increased risk of skin cancer. However, it is not biologically plausible that too much sun is associated with cavities. So you wouldn't even try to make that association, would you?

Public health researchers have had such a difficult time convincing people electromagnetic fields are harmful because, to date, there's been no convincing evidence the association is biologically plausible.

How do you get biological plausibility? Short of having lots of highly credible epidemiology, laboratory experiments with animals may be necessary. We'll talk more about this in the chapter on biological plausibility.

If you don't have any animal experiments, you're going to have to be creative... maybe even extremely creative. However, biological plausibility is not the equivalent of biological "truth" or "reality" and no one expects it to be (although you must proclaim it such). At best, it means a biological explanation that enjoys the firm possibility of a definite maybe.

Fortunately, it's likely no one will be able to prove you wrong. But your biological explanation should still pass the "red face" test. Depending, that is, on how confident you are no one will challenge you on this.

CHAPTER 3
The Significance of Significance

Assuming you've panned for a relative risk worthy of pursuit, you've got to ask yourself a question ... "Was I just lucky?" All your critics will ask that question, and, in this case, being lucky is not good.

You need to "prove" you weren't just lucky, that your relative risk was not a mere random occurrence, that it's real, that it's statistically significant. (Of course, if you can't do that, you can always try to ignore it. You might get away with it. In part, this will depend on how good of a job you did picking a risk to target.)

Statistical significance is an expression of how sure you are that your results did not occur by luck or chance, about how sure you are that they're not a fluke. Traditionally, conventionally and historically, a relative risk is statistically significant when we are 95 percent sure that it did not occur by chance.

Now, although the 95 percent level is not a law of nature and is not etched in stone, most scientists would be pretty embarrassed to label their results statistically significant at anything lower than 95 percent. In fact, many won't even consider publishing results that weren't statistically significant at a 95 percent level. You, however, cannot afford the luxury of being so finicky.

So, how do you perform the necessary statistical mumbo-jumbo to figure out whether your results are statistically significant at 95 percent? My advice is find some qualified statisticians and leave the divining to them. And here's what you do if your statisticians say your "p-value" is greater than 0.05 or your "95 percent confidence interval" includes a relative risk of 1.0 or less. Either means your relative risk is not statistically significant at 95 percent.

BORING TECHNICAL STUFF: The p-value indicates the probability that your statistical association (relative risk) is a fluke. The smaller number is better. A p-value of 0.05 or less means there is only a 5 percent or less chance that your relative risk is a fluke. You are halfway to achieving statistical significance at 95 percent.

Unfortunately, the p-value is immutable. Say your p-value is greater than 0.05 (very bad news). You're going to be stuck with it unless you (1) "adjust" your raw data or (2) ignore the p-value altogether. But if you play with the raw data, you'll probably have to recalculate your relative risk as well. In addition to a "good" p-value, you could also wind up with a higher (better) relative risk.

As for ignoring the p-value, most people don't even know it exists. Those who do know may think it's just another indecipherable statistical hieroglyphic. So if most people don't know or care, why worry about it?

The confidence interval represents the range of relative risks between which your risk number will likely fall. Let's say your statistician calculated a relative risk of 1.2 (a very weak association) and a 95 percent confidence interval of 1.05 to 1.35. Even though we calculated a relative risk of 1.2, we are almost 100 percent certain it does not equal the exact "true" risk. It's sort of a "best guess" of the true relative risk.

Even though we are nearly positive that 1.2 is not the precise and correct relative risk, we are 95 percent sure the "true" risk falls somewhere between the interval 1.05 to 1.35. All's well as long as the lower end of the risk range is above 1.0.

If the lower end of the range is above 1.0, then you can be at least 95 percent sure your relative risk is above 1.0. But if the lower end of the range is below 1.0, this means there is less than 95 percent confidence that your risk is real. This is bad. This means your relative risk is not statistically significant at a 95 percent level. What can you do about it?

If you are dealing with a lower range around 1.0, the simple solution is to narrow your confidence interval so that the lower end creeps above 1.0. How do you do that? You simply calculate a 90 percent confidence interval instead of the customary 95 percent confidence interval. Logically, less confidence means a narrower confidence interval while greater confidence means a wider confidence interval, and, after all, who needs greater confidence?

So you've lowered your standard for statistical significance from 95 percent to 90 percent. But you do have a defense in this case. Just say "Hey, 90, 95 percent, what's the difference? Close enough for government work."

You can try to "fix" your p-value or you can ignore it. You can adjust your confidence interval from 95 to 90 percent. Either way, it's sneaky. But remember: In risk assessment, it's not how you play the game; it's whether you win or lose.

CHAPTER 4
"Data" Collection

Back in the very first chapter, we said a good risk is ubiquitous and has exposures that are difficult to impossible to measure. Remember dioxin? How can you measure exposures that are found everywhere all the time? The advantage of exposures that are so difficult to measure is you don't have to "measure" them at all. You can just make 'em up.

But few people will ever notice or learn that your exposures aren't real. And it's almost impossible to validate your methodology or verify your results. So not only do you get to make up the data, but no one can check on you. Although this clearly violates one of the basic tenets of the scientific method (i.e., results should be capable of replication), it's a distinct advantage in the risk business.

Unfortunately, there are no general guidelines about how to do this. As in biological plausibility, you need to be creative. Maybe some of the examples we talk about here will inspire you.

Making the most of irrelevant documentation

Let's consider those epidemiologic studies that attempted to associate dioxin with cancer. Dioxin is ubiquitous in the environment; everybody is exposed to it and everybody carries some of it around in their fatty tissues. How do you study dioxin when everyone is exposed to it as part of everyday life? Isn't it impossible to find people who have not been exposed? Yes, absolutely. Has that stopped the risk assessors? Absolutely not!

Because dioxin is a by-product of certain industrial chemical processes, some epidemiologists theorized that chemical plant workers — by virtue of their employment status — should have relatively higher exposures to dioxin than the general public. So if dioxin is associated with cancer, there should be a higher incidence of cancer among these workers than among the general population. This theory almost makes sense.

So the epidemiologists decided to do a case-control study (the quick-n-dirty type, remember?) of the cancer rate of former chemical plant workers. Through employment records, our valiant epidemiologists chose subjects who worked in chemical processes in which it was assumed that some occupational exposure could have occurred. They also assumed the longer a worker worked in a process where dioxin exposure was possible, the greater the worker's exposure to dioxin.

The beauty of this scheme is that it is highly intuitive. It makes sense that chemical workers probably come into contact with chemicals. The longer they work with chemicals, the greater their exposure to these chemicals. This line of reasoning, however, says nothing about which specific chemicals the workers used, whether the chemicals included dioxin, whether dioxin was the only chemical the workers came into contact with or how much exposure to any chemical actually occurred.

Studies relying on this technique (which, by the way, have generally reported weak to very weak statistical associations) have been highly regarded by the public health community. Even though this ignores the metaphysical certainty that the chemical plant workers were exposed to many chemicals, not just dioxin. Or that it's impossible to tell from employment records if workers were even exposed to dioxin on the job.

From a purely scientific point of view, this data as well as these results mean absolutely nothing. But that's the beauty of this particular technique. It's so intuitive that it can overcome its own fatal shortcomings.

Thanks for the memories

Another useful data collection technique is the survey method, a technique that could be compared to political polling. It's not nearly as reliable, though. But it is very simple and straightforward. All you need is a series of "relevant" and preferably "loaded" questions that can be posed to a population of interest.

Typically, this involves asking your subjects whether they remember being exposed to the risk you're investigating. You may ask them to recall something from 50 years before. It's truly an amazing technique, considering people often can't remember what they had for dinner last week.

Sure, you must rely on the interviewees' memories of decades earlier. But typically, when you tell them what you're doing (trying to find someone to blame for their illness), they will be more than happy to remember whatever you want. Consider the diesel exhaust epidemiology studies.

Epidemiologists looking at diesel exhaust selected study subjects based solely on their status as workers who, at some time, either operated diesel equipment or worked in the vicinity of operating diesel equipment. For dioxin studies, researchers relied on employer-maintained employment records as the primary data source. But the diesel exhaust epidemiologists conducted personal interviews with workers, most of whom were elderly and many of whom had lung cancer.

This is masterful... let's see, I'm going to ask people dying of lung cancer if if they think diesel exhaust caused their illness. Do you think they'll try harder to remember a past exposure to explain the illness? It's called recall bias, it's pure genius, and it's guaranteed to give the "right" result.

The survey technique can even be taken one step further, if necessary. Maybe the person you want to interview is no longer alive. What should you do? The answer is simple: just interview the next of kin, like a son or daughter. Don't worry if they were not even around at the time of interest. Don't worry about using hearsay evidence. That's a trivial detail no one will ever get around to noticing anyway. You'd just be wasting good recall bias.

Unfortunately for the diesel exhaust epidemiologists, many independent and unbiased experts have since concluded the lack of definitive exposure data for diesel study populations precludes the use of available epidemiologic data to develop quantitative estimates of cancer risks. (You mean we can't estimate cancer risk for things just because we don't like them?)

I guess the dioxin epidemiologists got lucky.

CHAPTER 5
Mining for Statistical Associations

Once you've collected your data, how do you find the risk that's your ticket to stardom? There are two tried-and-true techniques virtually guaranteed to turn up something.

Disease Clusters and the Texas Sharpshooter

One of the best techniques is called the Texas Sharpshooter method. It goes something like this: The Texas Sharpshooter sprays the side of an abandoned barn with gunfire. He then draws a bull's eye target around a cluster of bullet holes that occurred randomly. He then can say, "See what a good shot I am!"

Basically, you can be your own sharpshooter if you find a cluster of disease and then shout "Aha!" or "Eureka!" or something to denote you've discovered the mother lode. Clusters are easy to find; they're everywhere, in fact. Epidemiologic studies of hazardous waste sites and electromagnetic fields are famous for clustering and the Texas Sharpshooter technique. For example, a study of a Woburn, Mass., site associated a cluster of 20 childhood leukemia cases with the site. It was very convincing. It didn't even matter that none of the contaminants at the site causes leukemia. That's the power of a cluster!

Consider, for example, one out of every three people in the United States will develop cancer sometime during their lifetimes. We call this the background risk or "natural" rate of cancer. It's yours by virtue of your birth. Now, if you do an analysis of cancer rates by geographic region or state or county or city or neighborhood, you will likely find that some areas will have a cancer rate of exactly 1 in 3. But most areas will have cancer rates that greater or less than 1 in 3.

Now, a real statistician will look at these rates and say, "Well, just by chance some areas will have higher cancer rates and some areas will have lower cancer rates. The differences average out as the geographic area gets larger. So the differences in rates between areas likely means nothing."

That may be, but you can't let that stop you. You've got to grab those areas with higher cancer rates and insist there's more to them than chance. Draw a bull's eye around the cluster you want and take it to the bank.

Data dredging: I know there's an association in here somewhere

Sometimes, clusters aren't obvious. You've got a river of data and nothing's making a ripple. What do you do? Well, what do you do when you're looking for something lost in a river? Simple, you get some dredging equipment and comb the river. By turning up everything, you hope you'll turn up what you want. Or if "you can't always get what you want, you just might find you get what you need" (Just kidding!). So what do you when you're looking for something lost in a river of data. *Data dredge!*

Conceptually, data dredging is like the Texas Sharpshooter technique except clusters are harder to find. You have to analyze your data forwards and backwards, from the top, bottom, and sides, from the inside out, and from the outside in. You slice it, dice it, and pick it apart any way you can to find an artifact (I mean risk) worth all this trouble.

All you need is a computer and a good statistical analysis program that can go through your data and look at every possible association. The computer does all the work — you get all the credit. All you have to do is pick the association you think makes your case and write it up. Let's look at a recent example.

A case-control study looked at risk factors for childhood leukemia, including environmental chemicals, electric and magnetic fields, past medical history, parental smoking and drug use, even dietary intake of certain food items. For "dietary intake of certain food items" alone, the study analyzed nine different foods, including breakfast meats, hot dogs, luncheon meats, hamburgers, charbroiled meats, oranges and orange juice, grapefruits and grapefruit juice, apple juice and cola drinks.

Obviously, right from the start, the researchers had no idea what they were looking for; they were simply on a fishing expedition. Amazingly enough, they caught a big one!

In examining the myriad of possible statistical associations, the study identified associations between a number of exposures and leukemia. These included breastfeeding, use of indoor pesticides, children's use of hair dryers, children's use of black-and-white television sets, incense use, father's occupation, mother's exposure to spray paints during pregnancy, other chemical exposures and home electrical wiring configurations. The association that received the most attention, however, was the one between hot dogs (eating more than 12 dogs a month, that is) and leukemia.

For this association, epidemiologists found a relative risk of 9.5, indicating, in their study, that children consuming more than 12 hot dogs per month were 9.5 times more likely to develop leukemia than children who consumed no hot dogs. The authors determined this association was biologically plausible because processed meats contain nitrites which may be precursors of other chemical compounds that have been associated with causing leukemia in rats and mice.

The researchers concluded their study "suggests" that diet is important to leukemia risk and that reduced consumption of hot dogs could reduce leukemia risk. A great result from a fishing expedition.

My only criticism is that the authors included in their writeup enough information for the careful reader to discern the study failed to come up with associations between other types of processed meats (including ham, bacon, sausage and luncheon meats) and leukemia. Given that these foods also contain nitrites and, therefore, should also be associated with leukemia risk, the authors should have omitted this information from their report. It only detracts from their conclusions about hot dogs.

CHAPTER 6
The Mixmaster Technique

What if you don't have the time or the money or the inclination to do your own epidemiologic study? What if others have already published epidemiologic studies on your risk but they didn't find anything convincing. Or some found something while others haven't? Well, just be very creative.

You could take the existing studies, assume they are similar enough to be combined and, *voila!*, you have an entirely new study. This technique is called meta-analysis. The best way to demonstrate the power of meta-analysis is to show you the greatest masterpiece, the Mona Lisa, of all meta-analyses: the Environmental Protection Agency's risk assessment on environmental tobacco smoke (ETS). There simply is no better example of this technique at work.

At the time the ETS risk assessment was conducted, there were 30 published (and who knows how many unpublished) epidemiologic studies on ETS conducted in a number of countries. Of the 30 published studies, eight reported statistically significant associations between exposure to ETS and lung cancer; 22 other studies reported either no association or no statistically significant association. Of the 11 studies that examined U.S. populations, only one reported a statistically significant association.

Realizing the difficulty of credibly associating ETS with lung cancer based on conflicting studies, the ever-resourceful EPA chose meta-analysis.

Using this technique, EPA combined the 11 U.S. ETS epidemiologic studies and came up with a relative risk of 1.19 that was statistically significant at a 90 percent confidence level (Note: Even though their results weren't statistically significant at a 95 percent level they were resourceful enough to claim statistical significance at a lower level. Another clutch decision!) With this "statistically significant" relative risk, EPA went on to estimate 3,000 lung cancer deaths can be attributed to ETS every year.

What's so amazing about all this? Well, EPA did such a good job picking a target for its risk assessment and meta-analysis that the intrinsic characteristics of the target itself were strong enough to overcome the scientific deficiency of the meta-analysis.

ETS was a classic target. The risk was unprovable (any risk would be too small to find, a fact borne out when 10 out of 11 U.S. studies turned up nothing). ETS is a common exposure. The cause-and-effect relationship in question is intuitive. The tobacco industry is easy to pick on. ETS is an involuntary risk. And, for non-smokers, there's no personal sacrifice involved in forcing others to quit. The technical deficiencies, while numerous and significant, were no match for these intrinsic characteristics.

Now remember, meta-analysis depends on the assumption the studies are similar enough to be combined. Yet mixing the different ETS studies is like mixing apples and oranges. You see, none of the ETS studies contains real exposure data. All the "exposure" data was derived from elderly women being prodded to remember their husbands' smoking habits of decades earlier (like the diesel exhaust studies). Or they came from the memories of other relatives.

None of this "exposure" data was ever validated or verified for accuracy. The clincher, however, is that each ETS study asked different types of study populations different questions about different time frames. To combine these studies together is truly the epidemiologic personification of the data processing acronym GIGO (garbage in, garbage out).

But, in the end, you've got to give credit where credit is due. EPA picked the right target — and hit the bull's eye. The rest is risk assessment history. Maybe this is really a lesson in picking a good target.

CHAPTER 7
Instant Risk

Haven't got time to do your own soup-to-nuts risk assessment? Then "instant" risk is for you. No fuss, no muss and guaranteed results. The classic example of this is risk assessment for ionizing radiation.

Everyone is exposed to ionizing radiation every day. It's unavoidable and natural. The two main sources of ionizing radiation are the earth and space. Soils and rock contain naturally occurring radioactive elements that either give off radiation or emit radioactive particles. Space is continually bombarding us with cosmic rays. You would not consider either of these to be dangerous because they occur naturally. Even if you lived the idyllic lifestyle in the Garden of Eden, you would still be exposed to ionizing radiation from these sources.

Some human populations have had very, very, very high exposures to ionizing radiation. Survivors of atomic bomb explosions. Uranium miners. Women who, in the 1920s, painted watch dials and instrument panels with radium paint and licked their brushes to get better points. Studies have shown a generally accepted association between these very, very, very high radiation exposures and cancer.

Notwithstanding what we know about high levels of ionizing radiation, there is not a generally accepted association between lower levels of ionizing radiation from manmade sources (like medical X-rays) or environmental levels of ionizing radiation from naturally occurring sources (like radon in the home).

Now ordinarily, you might conduct a case-control epidemiologic study to try to identify such an association and many folks have. But you don't *need* to. Just base your study on those of the atomic bomb survivors, underground uranium miners and radium watch dial painters, and you've got instant risk. How? Why?

Years ago, some genius came up with the theory that if something (say radiation) can be harmful at very high exposure levels, in the absence of knowledge to the contrary, it should be assumed it is harmful at any exposure level. This theory is known in risk assessment circles as the linear nonthreshold model.

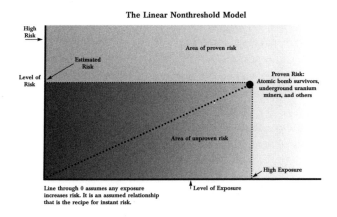

Using a graph similar to that above, all you need to do is measure or estimate the exposures to your population, find that exposure level on the graph and follow it over to a risk level. What could be easier? Just make believe that getting a medical X-ray is like surviving an atomic bomb explosion. Or that playing ping-pong in your basement rec room is like working in an underground uranium mine! Sounds silly, you say? Don't worry; this is one of the most commonly accepted tenets in the public health community.

You'll need to be prepared for real scientists who might say the linear nonthreshold model flies in the face of everything we know about risks from low levels of exposures. For example, studies of the atomic bomb survivors report an increased incidence of cancer only at the very highest exposures. Among those survivors with less than the highest levels of exposures, a decreased incidence of cancer (as compared to the general population) was observed.

Epidemiologic studies of workers show what is called the "healthy worker effect." That means despite being exposed to comparatively more "risks" on the job, workers are typically healthier than nonworkers. Finally, vaccines (like those for polio, measles, mumps, diphtheria and the like) intentionally expose humans to low levels of toxins but keep individuals healthy, not sick.

But, as I said earlier, the linear nonthreshold model is a public health mantra. It's not open to criticism.

A final word about the "instant risk" technique. It can save you lots of headaches. Consider the following story.

Not long ago, the National Cancer Institute conducted a very large and well-designed study to look at risk factors for lung cancer, including radon in the home. NCI's study failed to find an association between radon in the home and lung cancer. But at the same time, the Environmental Protection Agency was spending $20 million a year on its own radon program. When NCI published its results, EPA got upset.

Study results threatening the existence of the EPA's $20 million dollar radon program won't win friends or influence people in the program. They immediately screamed, "Fix this or else!" To atone for its sin, NCI repudiated its own epidemiologic study and published a new study applying the linear nonthreshold model to the underground uranium miner data. That produced an instantly acceptable risk assessment. And NCI and the EPA radon program were on speaking terms again.

The moral of the story? If you go nonlinear, you will be straightened out by your friends — *or else!*

CHAPTER 8
The Big Risk Number

You've calculated your relative risk and you've made it statistically significant. Is that enough? Can you just write up your results, get them published and start filling out those federal grant applications?

You can, but you haven't yet maximized your chances for success. There's one last thing to do and it's easy as pie. You simply take the innocuous relative risk number and "morph" it into a public health crisis.

You need to calculate a risk estimate for some population, preferably a large population or, better yet, all 250 million Americans. If you can figure the number of cancer cases or premature deaths associated with your risk, you're sure to get instant national attention. But how do you do this? Simple. Tell your statisticians you want to calculate an attributable risk. They know how.

Attributable risk is intended to indicate what percentage of deaths in a population are caused by a risk. For example, saying that "16 percent of all deaths are due to being overweight" is an attributable risk. You've attributed 16 percent of all deaths to obesity. All you need to do then is figure out how many deaths there are annually (about 2.2 million in the U.S., according to 1991 statistics), then multiply the number of annual deaths by the attributable risk (16 percent). *Voila!* A public health crisis is born!

EXAMPLES OF ATTRIBUTABLE RISKS	
Risk	**Annual Deaths Attributed to Risk**
Obesity	350,000 from all causes (Source: derived from 1995 Harvard University Study)
Smoking	390,000 from all causes (Source: U.S. Surgeon General)
Radon	40,000 from lung cancer (Source: U.S. EPA)
Chlorinated tap water	10,000 from bladder & rectal cancer (Source: Morris et al 1992)
Environmental tobacco smoke	3,000 from lung cancer (Source: U.S. EPA)

Now your statisticians (if they are competent and conscientious) should ask if you really want to calculate an attributable risk. This query will be based on the following warning that appears on the package of every statistical analysis program:

STATISTICIAN'S WARNING: ATTRIBUTABLE RISK MAY NOT BE SCIENTIFICALLY JUSTIFIABLE. IT IS CALCULATED FROM VERY UNCERTAIN STATISTICAL ASSOCIATIONS. THESE ASSOCIATIONS MAY NOT REFLECT TRUE BIOLOGICAL CAUSE-AND-EFFECT. AT BEST, A STATISTICAL ASSOCIATION IS A REPRESENTATION OF WHAT WAS OBSERVED IN A PARTICULAR POPULATION STUDIED AND IS NOT APPLICABLE TO OTHER POPULATIONS NOT STUDIED.

You, of course, should ignore this warning

CHAPTER 9
Cooking Up Biological Plausibility

Earlier we said biological plausibility is an essential element of the epidemiology equation. If your statistical association doesn't make some sort of biological sense, you're in trouble. For our purposes, biological plausibility can be developed in two ways: either from the linear nonthreshold model we discussed earlier or from toxicology experiments. We'll briefly discuss biological plausibility from the linear nonthreshold model and then spend most of this chapter on the technique of toxicology for risk assessment.

Instant biological plausibility

Biological plausibility from the linear nonthreshold model is simple. Whatever happens at high doses also happens at low doses. Cancer from too many medical X-rays is biologically plausible because of the studies of the A-bomb survivors. Forget that this is completely contradicted by those same studies. Skin cancer from low levels of arsenic is biologically plausible because high levels of arsenic are associated with skin cancer. Forget arsenic is actually an essential nutrient in trace amounts.

That any level of exposure to mercury is dangerous is biologically plausible because we know that high levels of exposure can lead to mercury poisoning. Never mind that mercury was used for more 350 years as an effective treatment for syphilis before it was replaced in the 20th Century by an arsenic compound.

Lung cancer from radon in the home is biologically plausible because of the observations from underground uranium miners. Don't let small details (such as areas with higher levels of radon are associated with lower lung cancer rates) bother you. Lung cancer from ETS is biologically plausible because smokers have a higher risk of lung cancer. Never mind that passive smokers are inhaling a chemically different type of smoke in a physically different manner. High exposures, low exposures, different exposures. They're all the same...

Baked biological plausibility

Toxicology is the study of poisons. It's not a traditional public health discipline mainly because poisons, as commonly thought of, are not a traditional public health issue. But so much for tradition. Today, toxicology is a multipurpose weapon in the public health arsenal.

Toxicology can be used to create a risk from scratch. For example, as you are reading this, the U.S. Government's National Toxicology Program is "investigating" whether alcohol causes cancer. There are only a few inconclusive epidemiologic studies for this association. So the government may have to "jump start" the research to go anywhere at all.

Toxicology can be used as a life support system to keep a risk alive. In the 1970s, toxicologists first reported dioxin is an extremely potent carcinogen in laboratory animals. However, epidemiology studies have since failed to associate dioxin with cancer in humans to any convincing degree. Nevertheless, the mere existence of the dioxin toxicology has fueled unending research (worth well over $1 billion) whether dioxin causes cancer and other health problems in humans.

Lastly, let's talk about toxicology's use for biological plausibility purposes. But there's one caveat: Even if you have the best toxicology in the world, few will believe you unless you've got human data that "proves" your point. That's why, for example, those who have been pushing the dioxin risk are still running in place after almost 20 years. The dioxin epidemiology is really disappointing, particularly considering how promising the dioxin toxicology was.

Most of the general public can distinguish — or are willing to distinguish — the difference between what happens to rats in "rigged" experiments and humans in the real world. So the dioxin toxicology itself hasn't been able to carry the day. Nevertheless, proponents of dioxin have obviously done quite well for themselves.

Being a public health-type, you probably don't know the first thing about toxicology, or certainly not enough to design and run your own bioassays (that's what this type of toxicology experiment is called). Unless you can find "canned" toxicology (studies that have already been conducted and published in the literature), you'll have to team up with a skilled toxicologist.

By skilled, I don't mean a toxicologist that is necessarily a "real" scientist — they're not inclined to engage in this sort of silliness...I mean, activity. You want someone who understands your need to find risk and who knows all the tricks to maximize your chances. You need someone who can pick the right animals, give them the right doses in the right way and who knows what to look for when the animals are autopsied.

Picking the right species

Since we don't typically study poisons directly in human subjects, we use rats, mice and other types laboratory animals. What species of animals should you use? You're probably going to need at least 300 animals (trust me). Generally speaking, rats and mice tend to be the most popular for bioassays because they're a lot cheaper, take up a lot less space, require a lot less care and are a lot less endearing than some other animals.

But remember: cutting corners may cost you in the long run. That is, the least expensive animals may also be the least sensitive to what you are exposing them. Not all animals react the same way to chemicals. What's toxic to a dog may have no effect — or a reduced effect — on a rat. So, instead of saving money by getting rats, you could actually waste your money by picking an animal that doesn't respond the way you want.

So which species do you pick? Unfortunately, there is no easy way to predict which species will give you the best results. It's basically bioassay roulette. If cost is not a showstopper, you could run your bioassay in several different species such as rats, dogs and rabbits. You then base your results on the species that is the most sensitive in the bioassay. The advantages of this are several.

First, at least you'll know you haven't picked the least sensitive species on which to base your results. Second, if the results of all the bioassays show all the different species get whatever you're looking for (*bravo!*), you have a self-validating series of interspecies experiments. On the other hand, if results between species groups are inconsistent (the rats get cancer but the dogs and rabbits don't), you may have to explain away the inconsistency. Although you probably could finesse this issue in your writeup, it's probably easier to ignore the data you don't want. My guess is you wouldn't be the first.

So remember, the most important thing is to find the proper species of test animal. This is the species most likely to help you "prove" your risk. Penny-wise is pound-foolish, as they say.

Maximizing the dose

Now that you've picked the right species of animal, how much do you expose them to? Simple: as much as they can stand. If your toxicologist were to expose the animals to the amount of your substance that humans ordinarily come into contact with, you would likely need, and I mean this literally, millions of animals... and millions of cages, tons of food and an army of lab assistants. This would probably blow your budget, even with one of those generous federal grants.

Such large numbers are needed because, even if the risk you're looking at really exists, it is likely to be so small it can only be detected by looking at a very large population. But, as you can see, large populations are impractical. Fortunately, there's a way around this dilemma; it's called the "maximum tolerated dose" or MTD for short.

The MTD is the highest amount of your substance that can be administered to the animals without killing them or making them noticeably ill just from being poisoned. But wait a minute, you say... Isn't that what we're trying to do anyway? Poison them? Well.. yes, but only in the long run. You don't want to just out-and-out poison them to death right away. Your results will be meaningless, and your competition...peers, I mean...will just laugh at you.

It's a well-established principle of toxicology that the dose makes the poison. In other words, everything is a poison; it's just a matter of dose. Chocolate ice cream can be a poison in high enough amounts. In trace amounts, arsenic is an essential nutrient. So you don't want to be accused of just "poisoning" your animals. The MTD basically maximizes the probability that a bioassay with only 50 animals will turn up something useful to you without simply poisoning the animals.

How high are MTD doses? MTDs can be anywhere from 10,000 times to 100,000 times what humans may be exposed to. But this has no relationship to reality, you say? Perhaps but remember, we're trying to manufacture risk and make you a star. Who's got time for common sense?

What to do with rats that just won't cooperate

Sometimes the logical or most obvious way to expose the laboratory animals to your substance won't do the trick. For example, if you want to look at dandruff shampoo as a cancer risk (and some have), you might consider applying the shampoo (or its key ingredient) to the skin of your laboratory animals.

Or, let's say you are interested in fiberglass as a cancer risk when inhaled. You might consider making your laboratory animals breathe air that contains fiberglass fibers. Now, suppose that in both these cases, the logical and intuitive routes of exposure produce nothing useful in the bioassays. What do you do?

Don't panic. *Don't ever panic.* There's almost always some way to fix things and get your intended results. If you can't go in through the front door, try the back door. If skin applications of dandruff shampoo don't work, try feeding. When inhaling fiberglass doesn't work, try injecting it into the lungs. If feeding doesn't work, try inhalation. There are a number of pathways into an animal. One of them should work, even if by mere chance! Don't worry about the relevance of the pathway. It's like horseshoes and hand grenades — close is plenty good enough.

What to look for

If you're interested in a cancer risk, when your toxicologist is looking at your animal's body parts, she's going to be looking for cancerous tumors. The more tumors the better. However, it may be that your animals have developed noncancerous (benign) tumors as well as cancerous ones. If your toxicologist only counts the cancerous tumors, there may not be enough tumors for you to identify a statistically significant risk.

So you tell your toxicologist to count both the cancerous and benign tumors. You see, in toxicology for public health, we simply pretend that benign tumors are really just like cancerous tumors. (Does this type of fantasy sound familiar? Remember the linear nonthreshold model?) It's like basing your batting average on how many times you hit the ball, fair or foul, not just on your base hits. A grounder or pop fly is as good as double. In this case, every tumor, cancerous or benign, is a home run!

Wrapping it up

Once your toxicologist counts the tumors, she'll tell you whether there are significantly more tumors in animals exposed to your substance than in the control group... almost exactly what we do in epidemiology studies. If so, you've got a positive association between your substance and cancer (in animals at least). This is just what you were looking for. This is good enough for biological plausibility.

If there's no positive association, you've got three options. First, you can go back to the drawing board and pick a new target. This will cost time and money. Second, you can rerun the bioassay, maybe trying a different species or a different exposure pathway. This will also cost time and money. Third, you can ignore the results of your bioassay and take your chances on the epidemiology alone. If you pick Option No. 3, your epidemiology must be extra-convincing to overcome your failure to produce on the toxicology side.

CHAPTER 10
Peer "Review"

Peer review is supposed to be technical review of your work by a small group of qualified experts, usually three to five in number. Generally, this happens before your writeup is published in a scientific journal. Typically, your reviewers will not have access to the original data. This is good. As a past editor of the *New England Journal of Medicine* once put it,

"If data have been cooked and the results plausible, there is no way peer review can catch the fraud."

    Arnold S. Relman, M.D.
    Editor-in-chief
    New England Journal of Medicine

    (from the *Washington Post* — May 16, 1989)

Nonetheless, if you've followed this guide in doing your research — and past is prologue — you shouldn't have any problem getting past peer review. Often, it's a mere formality. Be careful, though. It can also help prevent the inadvertent comment or study that otherwise causes the whole risk assessment house of cards to collapse.

Good peer review can prevent the sort of mistake the authors made in a study, recently published in *The New England Journal of Medicine*, that made national headlines. In this study, the authors associated obesity with premature death (Give them high marks for intuitiveness, low marks for sacrifice — Americans aren't ready to give up cheeseburgers, french fries or milk shakes).

The researchers conducted a cohort study (the better type of epidemiologic study that takes a long time to complete) involving 115,195 women free of cancer and cardiovascular disease when the study started in 1976. Results of this study are summarized below. They did such a good job "technically" that their results make no sense from a common sense perspective.

SUMMARY OF RESULTS FROM OBESITY STUDY			
	Underweight	Within 15% of desirable bodyweight	Overweight
Number of women	14,771 (13%)	31,992 (59%)	32,245 (28%)
Relative risk of early death for all women	1.0 (no increased risk)	0.8 (20% less risk)	1.05 (5% increase in risk)
Relative risk of death for nonsmoking women	1.0 (no increased risk)	1.05 (no increased risk)	1.43 (43% increase in risk
Relative risk of death for nonsmoking women with stable weight	1.0 (no increased risk)	1.2 (20% increase in risk)	1.7 (70% increase in risk)

For *all* women, including smokers and nonsmokers, the reported increased risk of premature death from being overweight is 5 percent. For *nonsmoking* women, the reported increased risk of premature death is 43 percent, and for *nonsmoking* women with stable body weight, the increased risk of premature death is 70 percent.

Gadzooks! So, if you're overweight, your risk of premature death is eight times greater if you're a nonsmoker, 15 times greater for nonsmokers who are old friends with their love handles. Can this be true? Will smoking become a health craze for the obese? Clearly, this result slipped by the peer reviewers. This is embarrassing.

So, peer review should be constructive. Keep a few thoughts in mind, however. First, to the extent you have the opportunity to pick your reviewers, pick ones you have confidence will help you out (not like the boneheads above). You want constructive criticism. By this, I mean criticism that will help you make even a more convincing case to the public, not criticism that will make your work more scientific — and less convincing.

You also want reviewers with stellar reputations whose credibility will rub off on your work. Fortunately, they are not hard to find. After all, the public health research community is small and tends to be protective of itself. Infighting and criticism about the credibility of research or researchers would only reflect poorly on the entire community. Worse, that could fuel public suspicions about public health research itself.

CHAPTER 11
The Final Document

The writeup of your research is crucial to success. Most important, you want to write a good abstract... it may be the only part of your writeup that many people will read.

In any case, the abstract, as well as your writeup, should exude confidence in the results described. Use words with strong, clear meanings. For example, say your research "proves" or "demonstrates" your conclusions or "confirms" the conclusions of others. You should avoid weaker terms like "indicates" or "evidences" or "supports the theory."

Importantly, your risk "causes" the disease of concern. Never say it is "associated" with the disease or "increases the risk" of the disease. It *causes* the disease. Never use the word "may" either. Although these weaker terms are more accurate from a purely scientific point of view, they are counterproductive when trying to communicate your risk to the general public. Qualifiers tell the public your results probably don't mean anything.

Always call your work "science" or "scientific research." That sounds better than what your research really is. DO NOT DESCRIBE THE SHORTCOMINGS OR LIMITATIONS OF YOUR RESEARCH. Even if you feel morally obligated to do so, for heaven's sake, don't include them in your abstract. Finally, the last paragraph of your writeup ought to clearly state how you think public policy should be changed because of your work. For example, say something like "This research demonstrates the need for the government to take immediate steps to ban the use of [your risk]." That's sure to draw attention.

Avoid being too subtle, like the following from a recently published study associating nitrous oxide (laughing gas) and miscarriage in dental assistants:

> "There is currently no mandatory Occupational Safety and Health Administration standard for nitrous oxide, although possible standards have been suggested...In the meantime, minimizing exposure seems prudent."

Be bold and straightforward. The statement above should read something like this:

> "Anesthetic gas causes dental assistants to miscarry. The Occupational Safety and Health Administration should ban its use."

And dental patients should just grin and bear it, I guess.

CHAPTER 12
Where to Publish

You should aim to get your research published in one of the better known scientific journals. These include *The New England Journal of Medicine* (the best known and most prestigious of all), *The Journal of the American Medical Association* (the AMA imprimatur is always very impressive), and *The American Journal of Public Health* (the official voice of the American Public Health Association and read by most card-carrying public health researchers).

Then there's the *Journal of the National Cancer Institute*, part of the federal government's famous (and cash-heavy) National Institutes of Health, and the *American Journal of Epidemiology*, published by the Johns Hopkins School of Hygiene and Public Health, the world's most prestigious school of public health.

All these journals, closely watched by the national press, are the most likely to provide you with instant national attention. They have the most prestigious editorial boards. They are the public health equivalent of "prime time" on network television.

There are other good, but lesser known, scientific journals out there. *Science*, published by the American Association for the Advancement of Science, for instance. But as the name implies, they might not be interested in the sort of work we're talking about here. Fortunately, there are many other publications out there. (Only a real public health geek knows how many.) In fact, there are so many, you are almost sure to get your results published somewhere.

If you're an "unknown," the prestigious journals may be somewhat out of reach. But, if you've followed this handbook and have a good story to tell, it's possible to overcome your... ahem... anonymity. You can still garner a substantial amount of attention with other journals, but you'll have to work a lot harder getting people to pay attention. Folks will wonder, "If your stuff is so good, how come it's not in a major journal?"

But if your research is published by a major journal, drop everything and make yourself completely available for interviews and media appearances. You may even want to think about media training courses to sharpen your communications skills. You've got to be able to tell a compelling story in a 20-second soundbite. Unfortunately, that's not something they teach in graduate school. But it may be more important than the research itself.

CHAPTER 13
Dealing with Criticism

You will undoubtedly encounter critics of your work. For the most part, these people are easy to deal with (that's why the risk business is booming). But you still need to be prepared.

Your best defense is a good offense. Put your critics on the defensive immediately. Call them "out of the mainstream of the scientific community." Or label them "hired guns for sale to the highest bidder." Make sure you mention you're only acting in the interest of "the public health" and that you're not personally profiting from your research. Most people don't think of research grants and the attendant perks as compensation.

Force your critics to respond to questions like "How else would you explain my results?" By the time they explain all about statistical associations, statistical significance, biological plausibility and the like, anyone who was listening will be asleep. These tactics should get you past the initial round of criticism at least.

For ensuing rounds, there are a couple of general rules to keep in mind. First, never debate a critic who has nothing to lose in the public health community. These people can be extremely dangerous. They *will* call a spade a spade, and you could wind up looking very foolish.

Now, there will probably be some scientists in the public health research community who will disagree with or be skeptical of your results; they may criticize your work openly, too. However, they will always be polite, careful not to attack you personally or the public health research community as a whole. But a critic who has no vested interest in the public health research community and who doesn't care if the whole system is scandalized should be avoided at all costs. You don't want to give such persons a forum. What if they actually get their message across? That could be catastrophic for your work and might even burst the risk assessment balloon itself. Your competition...er, peers would hate you for that.

Because you'll only debate people who are sensitive about maintaining the integrity of the system, never concede anything to them, even if you know you're dead wrong. As long as they play the game, they can't challenge you effectively.

Now that the basic tactical defenses are out of the way, let's talk about defenses for some of the substantive criticisms you may encounter.

If your key relative risk is a weak association (i.e., below 3), you'll get criticism. Why? Because, in reality, epidemiology is too crude a tool to find risks so small. As Dr. Charles H. Hennekens of the Harvard School of Public Health said recently:

Epidemiology is a crude and inexact science. Eighty percent of cases are almost all hypotheses. We tend to overstate findings, either because we want attention or more grant money.

(from the *New York Times* — October 11, 1995)

Remember how data was collected in the dioxin, environmental tobacco smoke and diesel exhaust epidemiologic studies? Didn't you think there was a lot of room for error? Particularly when you're looking for a very small increase in risk?

Consider the ETS case, for example. The annual background (or natural) rate of lung cancer among nonsmokers is 10 cases per 100,000 nonsmokers. EPA's ETS risk assessment says exposure to ETS increases this rate to a little less than 12 per 100,000 nonsmokers. Is it really possible to measure such a small increase in risk by asking small, nonrandom groups of people loaded questions about events 20, 30 or even 40 years ago?

You should ignore the part about epidemiology being a crude tool. How dare anyone cast aspersions about such a time-honored methodology? (Of course, it didn't get to be time-honored through this type of research!) Just say even small risks applied over a large enough population can be large risks.

This non sequitur isn't a fib. You just answered a different question. In effect, your critics asked something like "Aren't you trying to use the naked eye to see an atom?" You responded by saying, "Billions of atoms can constitute an object that is visible."

You may also be criticized for not fully addressing the possibility your risk was actually due to some factor that you have overlooked — or ignored! For example, in the case of the chemical plant workers and dioxin, there was a possibility (if not likelihood) that the workers came into routine contact with chemicals other than dioxin. These other chemicals, not dioxin, could have been responsible for any observed increase in cancer incidence.

Now that sounds like a tough one. How do you respond? Simple. Just give a lawyer-like answer: "There's no convincing evidence of any confounding risk factors..." Well of course there isn't. Either you were smart enough not to write about competing risk factors or you were smart enough not to collect any such information in the first place.

Then there's one last criticism: the results of other published studies conflict with yours in that they don't show an increase in risk. This is an easy one. All you say is "Negative data doesn't prove anything." If you have a million studies that show a risk doesn't exist — and I have *one* study showing the risk exists — your million studies are meaningless. Obviously this theory is true only in the most technical and absolute sense, but it has achieved the status of a commandment in the church of public health.

The likelihood of this type of criticism, however, is diminished by the phenomenon known as publication bias. Scientific journals, particularly those in public health, are extremely reluctant to waste space on research that doesn't show *something*. As a result, it's unlikely studies conflicting with yours will be published. And if they aren't published, they can't be used against you!

CHAPTER 14
A Final Word

All right, so it's almost 1996 and we're about 20 years into the public health gold rush. How much longer will it last? How much longer before the general public realizes Americans are healthier and living longer than ever before? Well, past is prologue, so I'm confident the public health gold rush is far from over.

Consider alchemy, the ancient art that sought to transmute base metals, such as lead, into silver and gold. It was the forerunner of modern chemistry — and maybe even meta-analysis? Alchemy appeared back in the 5th century B.C. and lasted well into the Middle Ages. It didn't fall into disrepute until some alchemists became obsessed with a quest for the secrets of transmutation and adopted deceptive methods of experimentation. Hmmm... you don't think... nah, it'll take forever for the public to figure out the shady type of risk assessment.

Then there's astrology. Astrology is the practice of foreseeing future events through omens or signs. (Do you think a statistical association is an omen?) It's based on the theory that the movements of celestial bodies influence human affairs and the course of events. Astrology was first practiced by the Assyrians (around the first millennium B.C.) and continued as a serious form of study into the 17th century when Christian theologians waged all-out war against it. Although their work eventually helped undermine it, the most famous early scientists, including Copernicus, Tyco Brahe, Johannes Kepler, Galileo, René Descartes and Isaac Newton, were all practicing astrologers.

In contrast, science, as defined by the scientific method, is still relatively young. Using this discipline, we learn things about our world slowly, over time, through piece-by-piece observation and experimentation. Science is not a quick-n-dirty, one-study endeavor. It has served mankind well over time, but we've seen fit to shunt it to the side, particularly recently, in the name of public health. Fortunately for the ambitious, this trend does not appear likely to change anytime soon.

So the big question is: "Will this guide be of any practical use in the future?" The answer, unfortunately, is "Yes, and for years to come."

Lexicon

Attributable risk *n.* The big risk number that puts you in the national spotlight.

Background risk *n.* The risk caused by birth. Currently estimated to be 1 in 3 for cancer. To be ignored because it likely dwarfs any risk you could possible find.

Benign tumor *n.* A noncancerous tumor that toxicologists pretend is cancerous.

Bioassay *n.* A laboratory experiment in which the risk of premature mortality for the experimental animals is 100 percent.

Biological plausibility *n.* A fantasy ostensibly based in biology that explains why a statistical association represents real risk.

Case-control study *n.* A type of epidemiologic study conducted by type A personalities... those who need results NOW!

Cohort study *n.* A type of epidemiologic study conducted by type B personalities... those who have got lots of time to kill, maybe 20 years or more.

Confounding risk factor *n.* A competing risk factor to ignore or belittle.

Data dredge *v.* To analyze a set of epidemiologic data every conceivable way to identify a positive statistical association between an exposure and a disease. Takes advantage of the fact that if enough analyses are conducted, a positive statistical association will turn up by sheer chance.

Epidemiology *n.* A methodology in which randomly occurring statistical associations are elevated to a cause-and-effect status.

Linear nonthreshold model *n.* A risk assessment theory that assumes if something poses a risk at high levels of exposure there is a risk at any level. Why risk assessors compare a dental X-ray to Hiroshima.

Meta-analysis *n.* An epidemiologic technique for turning a lot of nothing into something. compar. alchemy.

Negative data *n.* Conflicting research you'll need to explain away, preferably by belittling it.

Peer review *n.* Pre-publication congratulations from your colleagues. *obs.* An impartial, pre-publication review of the scientific merits of research.

Public health *n.* What we have now that the average American's life span exceeds 75 years. orig. What they didn't have in the 14th Century when one-third of all Europeans died from bubonic plague. Profession responsible for ensuring the U.S. public doesn't connect these facts.

Publication bias *n.* The tendency in the public health community not to publish studies inconsistent with the goal of finding risks.

Recall bias *n.* A phenomenon of case-control studies where subjects remember what epidemiologists want them to remember — even if it never happened.

Relative risk *n.* A type of statistical association routinely used in epidemiology. Actually has nothing to do with health risk. What you get when you merge a statistical association and ambition.

Risk assessment *n.* A recipe for success in public health. *obs.* a four-step process to identify and quantify hazards to human health.

Statistical association *n.* A highly prized numerical relationship between the exposure you're interested in and a disease. An omen of success.

Statistically significant *adj.* Confidence that a statistical association did not occur by chance. The standard level of confidence is 95 percent, but the ambitious may resort to lower levels of confidence (like 90 percent).

Strong association *n.* A relative risk greater than 3.0. The Holy Grail of public health research.

Toxicology *n.* The study of poisons. In risk assessment, a process where animals are poisoned and then sacrificed to see if the poison worked.

Weak association *n.* A relative risk between 1.0 and 3.0. Requires a snow job of unbelievable proportions to sell this as a genuine risk.

ABOUT THE AUTHOR

Steven J. Milloy is Director of Science Policy Studies at the National Environmental Policy Institute.

Since 1990, Mr. Milloy has focused exclusively on environmental, risk and regulatory policy issues. In 1993, Mr. Milloy led a study examining the relative roles of science and policy in regulatory risk assessment for the Department of Energy. This study resulted in the publication of *Choices in Risk Assessment: The Role of Science Policy in the Environmental Risk Management Process* in 1994.

Mr. Milloy is also the author of *Science-Based Risk Assessment: A Piece of the Superfund Puzzle*, a study of how risk assessment has been, and should be, used in the Superfund hazardous waste cleanup program.

Mr. Milloy has testified before Congress about risk assessment and Superfund. He has lectured on risk assessment at the Department of Energy, the Environmental Protection Agency, the Defense Department's Industrial War College and Uniformed Services University of the Health Sciences, the National Institute of Medicine, The Johns Hopkins University, The Catholic University of America, the Manhattan Institute and the American Policy Center.

Mr. Milloy holds a B.A. in natural sciences from The Johns Hopkins University, a Master of Health Sciences in biostatistics from The Johns Hopkins University School of Public Health, a Juris Doctorate from the University of Baltimore and a Master of Laws in securities regulation from the Georgetown University Law Center.

CATO INSTITUTE

Founded in 1977, the Cato Institute is a public policy research foundation dedicated to broadening the parameters of policy debate to allow consideration of more options that are consistent with the traditional American principles of limited government, individual liberty, and peace. To that end, the Institute strives to achieve greater involvement of the intelligent, concerned lay public in questions of policy and the proper role of government.

The Institute is named for Cato's Letters, libertarian pamphlets that were widely read in the American Colonies in the early 18th century and played a major role in laying the philosophical foundation for the American Revolution.

Despite the achievement of the nation's Founders, today virtually no aspect of life is free from government encroachment. A pervasive intolerance for individual rights is shown by government's arbitrary intrusions into private economic transactions and its disregard for civil liberties.

To counter that trend, the Cato Institute undertakes an extensive publicaitons program that addresses the complete spectrum of policy issures. Books, monographs, and shorter studies are commissioned to examine the federal budget, Social Security, regulation, military spending, international trade, and myriad other issues. Major policy conferences are held throughtout the year, from which papers are published thrice yearly in the *Cato Journal.* The Institute also publishes the quarterly magazine *Regulation.*

In order to maintain its independence, the Cato Institute accepts no government funding. Contributions are received from foundations, corporations, and individuals, and other revenue is generated from the sale of publications. The Institute is a nonprofit, tax-exempt, educational foundation under Section 501(c)3 of the Internal Revenue Code.

CATO INSTITUTE
1000 Massachusetts Avenue, N.W.
Washington, D.C. 20001

DATE DUE

2/21/07
ZTM
3/9/05